Creatures of Legend

CHUPACABRA

LATIN AMERICA'S BLOODTHIRSTY PREDATOR

DiscoverRoo
An Imprint of Pop!
popbooksonline.com

by Elizabeth Andrews

abdobooks.com

Published by Pop!, a division of ABDO, PO Box 398166, Minneapolis, Minnesota 55439.

Printed in the United States of America, North Mankato, Minnesota.

052022
092022

THIS BOOK CONTAINS RECYCLED MATERIALS

Cover Photo: Shutterstock Images

Interior Photos: Shutterstock Images, Jeff Carter/HowStuffWorks, Alamy Stock Images, Getty Images

Editor: Bridget O'Brien
Series Designer: Laura Graphenteen

Library of Congress Control Number: 2021951855

Publisher's Cataloging-in-Publication Data

Names: Andrews, Elizabeth, author.

Title: Chupacabra: Latin America's bloodthirsty predator / by Elizabeth Andrews

Other title: Latin America's bloodthirsty predator

Description: Minneapolis, Minnesota : Pop, 2023 | Series: Creatures of legend | Includes online resources and index

Identifiers: ISBN 9781098242336 (lib. bdg.) | ISBN 9781098243036 (ebook)

Subjects: LCSH: Chupacabras--Juvenile literature. | Goatsucker--Juvenile literature. | Animals--Folklore--Juvenile literature. | Fabled creatures--Juvenile literature.

Classification: DDC 001.944--dc23

WELCOME TO DiscoverRoo!

Pop open this book and you'll find QR codes loaded with information, so you can learn even more!

Scan this code* and others like it while you read, or visit the website below to make this book pop!

popbooksonline.com/chupacabra

*Scanning QR codes requires a web-enabled smart device with a QR code reader app and a camera.

TABLE OF CONTENTS

CHAPTER 1

OUT FOR BLOOD

No cow, goat, or sheep is safe when the Chupacabra is on the prowl. Known for draining livestock of blood, the Chupacabra wanders the lands of Latin America and the southern United States. Farmers fear it. Children hunt it. No one knows what it is though.

WATCH A VIDEO HERE!

Reports of the Chupacabra made farmers fear for their animals' survival every night.

People have reported seeing two different kinds of Chupacabras. The first is about four feet (1.2m) tall. It looks like a **reptilian** kangaroo. It has sharp spikes running down its back. This Chupacabra moves around by hopping on its back legs. All sightings of the reptilian

The Chupacabra's eyes are said to glow.

Chupacabra were in Puerto Rico. The other kind of Chupacabra is smaller and doglike. It walks on all four legs and is hairless with sharp teeth.

People consider these creatures to be the same monster because they leave the same destruction in their paths. Farmers awake to find their animals lying dead in their pastures completely drained of blood. Nothing but **puncture** wounds to their chests could explain

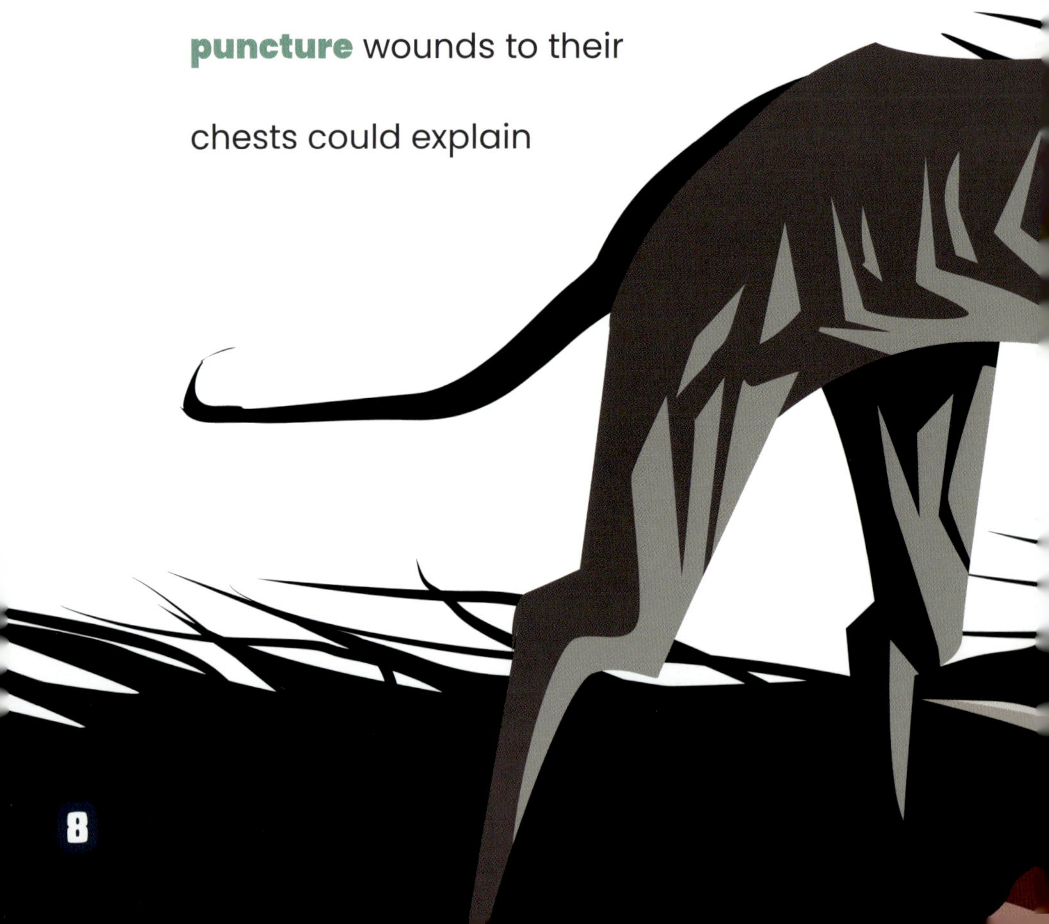

the cause of death. The strange attack is what gave the Chupacabra its name. *Chupar* means "to suck" and *cabra* means "goat" in Spanish.

Chupacabras have long, sharp teeth.

CHAPTER 2

THE ORIGINAL MONSTER

Puerto Rico is an island in the Caribbean Sea. The people of Puerto Rico are a blend of **indigenous**, African, and European cultures. This mix is common for countries in Latin America. Stories of

DID YOU KNOW?

Indigenous Puerto Ricans are called *Taínos*, meaning "men of the good."

Taíno people worshiped more than one god.

legendary gods and creatures from long ago have been passed down through word of mouth.

LEARN MORE HERE!

Chupacabras are usually three to four feet (.9-1.2m) tall.

Many of the original folktales talked about scary monsters and demons who roamed the island after dark. With thick forests that could hide almost anything, the land of Puerto Rico is a perfect place for spooky legends to be born.

The first reported Chupacabra sighting was in the spring of 1995. A girl named Madeline was looking out her window when the creature hopped across her yard. Fearing for her family's land and animals, Madeline ran out toward the monster and scared it away.

A neighbor of Madeline's tackled the Chupacabra at the edge of the woods. The neighbor could only hold onto the monster for a moment, but he saw that it had red eyes and razor-sharp fangs.

Farmers didn't want the Chupacabra anywhere near their land.

Soon, other farms began losing animals in the same way. Puerto Rico didn't have any large predators that could cause that kind of harm. The country became terrified of the monster. More sightings occurred and eventually the government got involved.

DID YOU KNOW?

People of Puerto Rico think the god *Yúcahu* lives in the forest. Some make sacrifices to him for his favor and protection.

In the fall of 1995, the mayor organized a hunt. Over 200 people volunteered to go into El Yunque, Puerto Rico's national rain forest. It was the perfect place for a monster to hide.

Unfortunately, the hunt was unsuccessful. After barely beginning, lights flashed throughout the forest and scared people away. The mayor went on an international TV show to talk about the monster haunting his town.

The people of Puerto Rico were afraid of what could be hiding in El Yunque.

CHAPTER 3

SMALLER, BUT STILL SCARY

After the show, more reports started flooding in. People from outside of Puerto Rico began being **terrorized** by the creature. Many of the sightings spoke of more animals drained of their blood. By 1996 over 1,000 animal deaths were blamed on the Chupacabra.

EXPLORE LINKS HERE!

Jorge Talevera from Nicaragua started waking up to dead livestock in 2000. He chose to **stake out** his fields hoping to catch whatever creature was responsible. During the stake out, Talevera saw three smaller monsters **stalking** his goats. They had bat-like skin and spikes running down their backs.

Farmers feared Chupacabras because they didn't know when the creatures would strike next.

DID YOU KNOW?

Some people believed that the Chupacabra was an alien that escaped from government testing.

Days later, Talevera came across the body of one of the monsters. He called a veterinarian to look at the body. The vet studied it and explained that it was a genetically engineered animal. This means scientists created it. The story fit **theories** that were already being spread. People had been saying the Chupacabra was a government experiment gone wrong.

The monster's body was sent to a university for more testing. This time, the results said it was just a dead dog. But when Talevera got the body back, it had been switched! He became suspicious of what he was being told.

The Xoloitzcuintli is known as the Mexican hairless dog. People confused them for Chupacabras.

MRS. CHUPACABRA

Phylis Canion lives in Texas on a ranch. In 2007, her chickens started showing up drained of blood. She reportedly found three Chupacabra bodies days later. After testing, they were proved to be coyotes. Canion insists the bodies she has are the monster. People from all over contact Canion when they see Chupacabra-like creatures because they think she's an expert. She's even been in documentaries!

However, the report Talevera got matched what was occurring in the United States and Mexico at the time. People were finding smaller versions of the Chupacabra on their land. Some were able to trap these little monsters and send them to scientists for testing as well. Like

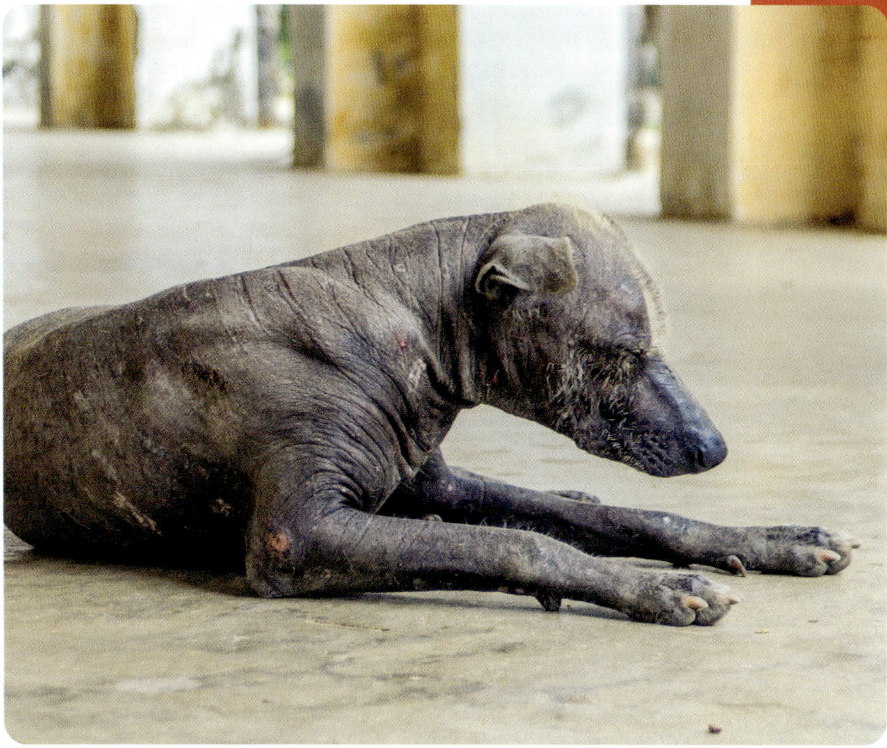

Mange spreads easily between animals.

Talevera's creature, they turned out to be dogs or coyotes. The reason they looked so frightening was because they were sick with **mange**.

CHAPTER 4

WHERE DID IT COME FROM?

A long time ago, people feared the natural world. They didn't understand why things like hurricanes, earthquakes, and wildfires happened. Mythical creatures were invented to bring some answers to these chaotic and damaging events. It

COMPLETE AN ACTIVITY HERE!

A statue of the Taíno moon goddess.

comforted people to think that a storm was caused by an angry god, rather than something they could not understand.

DID YOU KNOW?

A movie called *Species* came out in 1995. The monster looks like how the original Chupacabra was described. Some think it affected what people believed they were seeing.

In 1995, when people began seeing the Chupacabra, Puerto Rico and the United States had a rocky relationship. Puerto Ricans did not trust the US government. They were not offered the same rights as American citizens. When livestock started dying and negatively affecting Puerto Ricans' daily lives, it was easy to place the blame on the US government.

Some Chupacabra characteristics make it seem like an alien.

One of the main explanations for the Chupacabra's existence was that it was a government experiment gone wrong. The experiment got out and started wreaking **havoc** on the island. At the time, technology and medicine were advancing. Life was changing along

People sometimes fear what could happen if a scientific experiment goes wrong.

with it. People were uncomfortable with how quickly their world was changing and they began to fear it. Farmers losing their livestock were quick to agree that the advancements had something to do with the monsters.

The mystery of the Chupacabra has never been solved in Puerto Rico. As late as 2019, people on the island are still reporting attacks on their livestock. However, the Chupacabras that were seen in Mexico and the southern United States were proven to be coyotes many times.

MAKING CONNECTIONS

TEXT-TO-SELF

Do you think there was ever a real Chupacabra terrorizing farmers? Explain why or why not.

TEXT-TO-TEXT

Have you read any other books about legendary creatures? If so, what did they have in common with the Chupacabra?

TEXT-TO-WORLD

Why do you think people have fascinations with monsters?

GLOSSARY

havoc — wide and general destruction.

indigenous — the earliest known inhabitants of a place.

mange — a skin disease that causes hair loss.

puncture — a hole or wound.

reptilian — resembling a reptile.

stake out — to continuously watch a place or person.

stalking — tracking and following.

terrorize — cause to feel terror or extreme fear.

theory — an explanation of how or why something happens.

INDEX

ONLINE RESOURCES

popbooksonline.com

Scan this code* and others like it while you read, or visit the website below to make this book pop!

popbooksonline.com/chupacabra

*Scanning QR codes requires a web-enabled smart device with a QR code reader app and a camera.